TELL ME HOW IT WORKS

HOW DO WIND TURBINES WORK?

KATE MIKOLEY

PowerKiDS press
New York

Published in 2021 by The Rosen Publishing Group, Inc.
29 East 21st Street, New York, NY 10010

Copyright © 2021 by The Rosen Publishing Group, Inc.

All rights reserved. No part of this book may be reproduced in any form without permission in writing from the publisher, except by a reviewer.

First Edition

Editor: Siyavush Saidian
Book Design: Reann Nye

Photo Credits: Cover visdia/Shutterstock.com; Series Art (gears) goodwin_x/Shutterstock.com; Series Art (newspaper) Here/Shutterstock.com; p. 5 Mimadeo/Shutterstock.com; p. 6 Teun van den Dries /Shutterstock.com; p. 7 ZU_09/E+/Getty Images; p. 9 SkyLynx/Shutterstock.com; p. 11 TimSiegert-batcam/Shutterstock.com; p. 12 Mike Kemp/Getty Images; p. 15 Mischa Keijser/Shutterstock.com; p. 16 ilbusca/E+/Getty Images; p. 17 Joseph Sohm/Shutterstock.com; p. 19 (top) XXLPhoto/Shutterstock.com; p. 19 (bottom) picture alliance/Getty Images; p. 21 Oleksii Sidorov/Shutterstock.com; p. 22 Terry Eggers/The Image Bank/Getty Images.

Cataloging-in-Publication Data

Names: Mikoley, Kate.
Title: How do wind turbines work? / Kate Mikoley.
Description: New York : PowerKids Press, 2021. | Series: Tell me how it works | Includes glossary and index.
Identifiers: ISBN 9781725318298 (pbk.) | ISBN 9781725318311 (library bound) | ISBN 9781725318304 (6pack)
Subjects: LCSH: Wind turbines–Juvenile literature. | Wind power–Juvenile literature.
Classification: LCC TJ828.M55 2021 | DDC 333.9'2–dc23

Manufactured in the United States of America

CPSIA Compliance Information: Batch #CWPK20. For Further Information contact Rosen Publishing, New York, New York at 1-800-237-9932.

CONTENTS

WORKING WITH WIND 4
IT'S KINETIC! 6
TOWERING TURBINES 8
PARTS OF THE TURBINE 10
TURBINES OF ALL KINDS 14
GETTING ENERGY OUT 18
MAKING ENERGY CLEANER 20
POWERED BY WIND 22
GLOSSARY 23
INDEX 24
WEBSITES 24

WORKING WITH WIND

Have you ever thought about where electricity comes from? When you need to power a game or a tool, you plug it into the wall and it works. But how does that happen?

In the United States, most electricity is made by burning **fossil fuels**. These are nonrenewable. This means once they run out, they're gone forever. Fossil fuels are also harmful to the **environment**.

For years, scientists have been working on better ways to make electricity. Wind turbines are one method they've come up with.

About 64 percent of electricity used in the United States came from fossil fuels in 2018. Only 6.5 percent came from wind.

IT'S KINETIC!

Kinetic energy is a kind of power that comes from something being in motion. Wind has kinetic energy. This energy can be turned into electricity. To do this, a machine that can **convert** the energy is needed. This is where wind turbines come in.

GENERATOR

TECH TALK

There's a part inside a turbine called a generator. This is the part of the machine that actually makes the electricity.

Each year since 2005, 3,000 new wind turbines have been built in the United States on average.

A turbine is an engine or motor with blades that move with the help of water, steam, or air. The movement of air from wind makes the blades on a wind turbine move.

TOWERING TURBINES

When many wind turbines are grouped together in an area, it's called a wind farm. The electricity from one wind farm can power thousands of homes! There are a few kinds of wind turbines. The ones you're most likely to see on a wind farm are called horizontal-axis wind turbines (HAWTs).

HAWTs have two or three blades. The blades are attached to the **rotor**. When the wind blows, both the blades and the rotor spin. A long, pole-like structure, called the tower, supports the turbine.

TECH TALK
The higher up you go, the faster the wind gets. For this reason, taller towers help wind turbines take in more energy.

Some of the largest turbines have towers more than 780 feet (237.7 m) high and blades more than 350 feet (106.7 m) long!

PARTS OF THE TURBINE

A turbine's rotor connects to its generator, usually by a part called a shaft. The generator and shaft are both inside a larger piece called the nacelle, located at the top of the tower. The nacelle holds many of the parts that help make the wind turbine work.

As the wind on the outside turns the blades, the parts on the inside work together to use its power. The generator spins and turns the energy from this movement into electricity.

TECH TALK

Turbines only work with certain wind speeds. If there's no wind, there's no energy to use. However, too much wind can **damage** the turbine.

A piece called the anemometer is outside the nacelle. It measures wind speed and sends the data to a part inside called the controller, which turns the machine on or off.

Wind doesn't always move in the same direction. Some kinds of turbines won't work if the wind isn't blowing the right way. Instead, they need to be turned so they're facing the right way when the wind changes its direction.

There's a part called the yaw drive near the top of some turbine towers. It moves the turbine to make sure it's always facing the wind. A part called the yaw motor powers the yaw drive.

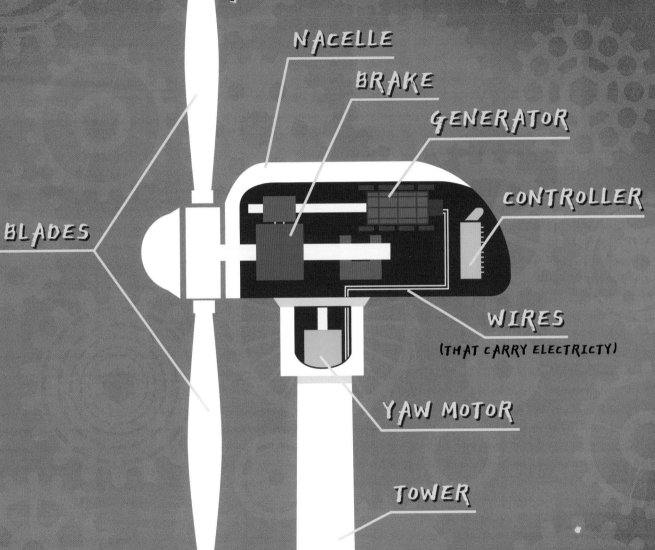

Inside the nacelle, there's a brake that can stop the rotor in **emergencies**.

TURBINES OF ALL KINDS

Wind turbines come in many shapes and sizes. Some families have small ones that power their homes. Wind farms with larger turbines can power many homes or businesses.

Offshore wind turbines are built on bodies of water, commonly the ocean. Wind is typically stronger over water off a coast, making these areas a perfect environment for turbines.

In shallow water, turbines can be built into the ground. **Engineers** are working on ways to make **platforms** that can float and hold turbines in deeper water.

TECH TALK

Since there's so much wind at sea, turbines can be made larger to produce more electricity. However, offshore turbines are harder to get to when they need to be fixed.

Some offshore wind turbines have blades as long as a football field!

Another kind of wind turbine is called a vertical-axis wind turbine (VAWT). These are less common than HAWTs. The blades on these turbines commonly spin around a vertical pole. Because of the direction of the blades and the way they move, VAWTs don't need to be moved to face the wind. They can use wind coming from any direction.

VAWTs come in a few different shapes. Engineers are still working to find the best design.

The VAWTs available right now don't work as well for the cost as HAWTs do. However, engineers are always working on improving **technology**.

17

GETTING ENERGY OUT

Now you know how wind turbines make electricity. But how does that electricity get distributed, or sent out, to the places it needs to power?

On a wind farm, the turbines are connected to something called the power grid. The power grid uses many power lines to move electricity to different areas. These lines carry electricity at a very high **voltage**. By the time it reaches a home or other building, the electricity has been converted to a usable voltage.

TECH TALK

The process of moving electricity from the place it was produced to the place it powers is called transmission.

You may be able to see power lines on some wind farms. However, these lines are sometimes laid underground.

MAKING ENERGY CLEANER

Compared to burning fossil fuels, using wind turbines is a cleaner, renewable way to get electricity. Fossil fuels put harmful matter into the air and pollute the environment. When they're working properly, wind turbines don't do this.

However, there have been some rare cases of liquids from turbines leaking. This isn't common, but when it happens, it does cause pollution. Another complaint people have about wind turbines is that they can kill birds and bats that fly into them.

Scientists are working on ways to make wind turbines even cleaner and better for the environment than they already are.

POWERED BY WIND

Plenty of **resources** on our planet could run out, including fossil fuels. Wind, however, isn't going anywhere. As long as there's wind, wind turbines can be used to make electricity.

Now that you know how they work, ask a parent or teacher if there are any wind farms in your area. While wind turbines still aren't used as much as other energy sources, they're becoming more widespread. Are there places in your community powered by wind?

GLOSSARY

convert: To cause to change form.

damage: Harm. Also, to cause harm.

emergency: An unexpected situation that needs quick action.

engineer: Someone who plans and builds machines.

environment: The natural world.

fossil fuel: Matter formed over millions of years from plant and animal remains that is burned for power.

platform: A raised structure with a flat surface where people or machines do work.

resource: A usable supply of something.

rotor: A part of a machine that turns around a central point.

technology: Tools, machines, or ways to do things that use the latest discoveries to fix problems or meet needs.

voltage: A measurement of electrical energy.

INDEX

A
anemometer, 11

B
bats, 20
birds, 20
blades, 7, 8, 9, 10, 15, 16
brake, 13

E
electricity, 4, 5, 6, 8, 10, 14, 18, 20, 22

G
generator, 6, 10

H
horizontal-axis wind turbine (HAWT), 8, 16, 17

K
kinetic energy, 6

N
nacelle, 10, 11, 13

O
offshore wind turbine, 15

P
pollution, 20
power grid, 18

T
tower, 8, 9, 10, 12

V
vertical-axis wind turbine (VAWT), 16, 17

W
wind farms, 8, 14, 18, 19, 22

Due to the changing nature of Internet links, PowerKids Press has developed an online list of websites related to the subject of this book. This site is updated regularly. Please use this link to access the list: www.powerkidslinks.com/tmhiw/turbines